YOUR BODY'S HEROES AND VILLAINS

MICROEXPLORERS

First edition for the United States and Canada
published exclusively 1997 by Barron's Educational Series, Inc.

Originally published in English under the title
Your Body's Heroes and Villains © Copyright Useful Books, S.L.,
1997 Barcelona, Spain.

Authors: Norbert Landa and Patrick A. Baeuerle
Illustrators: Antonio Muñoz, Ali Garousi and Roser Rius
Graphic Design: IGS - Barcelona, Spain

Address all inquiries to:
Barron's Educational Series, Inc.
250 Wireless Boulevard
Hauppauge, NY 11788

Library of Congress Catalog Card No. 97-74826

International Standard Book Number 0-7641-5051-0

Printed in Spain
9 8 7 6 5 4 3 2 1

YOUR BODY'S HEROES AND VILLAINS

MICROEXPLORERS

Learning about immune cells: the tiny defenders that safeguard our lives against nasty bacteria and viruses

Norbert Landa and Patrick A. Baeuerle

BARRON'S

Welcome to the

Everybody knows how it feels to have a cold. Our nose is stuffed or runny, and we feel sick and tired. We have to sneeze. Atchooo!

We know all about being sick. We even know that tiny germs cause colds and influenza, and that people can infect each other. Germs are also called microbes.

How do germs make us sick? How do they look? What exactly do they do? How does our body protect itself against them? How do we get rid of them? What really happens when we have a cold?

Unfortunately, we are much too big to see all those tiny microbes and cells working and fighting inside our body. Our eyes are not able to see those fascinating struggles. With one single sneeze, millions of them land in the tissue. All we see of them is—you guessed it—mucus.

tour!

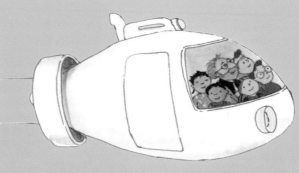

Let's have a closer look at that. Well, let's not look so much at mucus. Instead, let's watch the incredibly tiny villains causing colds and many other serious diseases, and our body's heroes that are fighting them.

I invite you on an expedition with our MicroMachine. It will shrink us to the sizes of the invaders and defenders. The villains are bacteria and viruses, and the heroes are our immune cells. I, by the way, am Professor Gene, your tour guide.

Actually, this topic is a matter of life and death. In each and every moment of our lives, and not only when we feel sick, various kinds of microbes are trying to invade us. They want to make their living within the cells of our body because, to microbes, our body is a warm, pleasant, and nourishing environment. Let's find out about the dirty tricks microbes play. Let's learn about the clever counterstrategies our defender cells use to keep our body clear and healthy. Let's investigate why microbes can make us ill. Let's see what really happens at an inflammation site, why fever can be quite helpful sometimes, and how antibiotics and vaccinations can support our defender cells.

Within a nose infected by cold viruses, a battle is occurring. Let's start to shrink and look into it!

About people

We human beings share this world with microbes, which are tiny living germs. Bacteria, viruses, fungi, and parasites live in places that are comfortable, where they can eat and multiply. They live everywhere— in the earth, in the seas, in the air we breathe, and in the food we eat. They live in and on other beings. As many microbes (mostly harmless) are on our skin as there are people on our planet. There are as many kinds of microbes as animals in the world.

See this bizarre landscape? Actually, this is the surface of a tooth we are drifting over. It is covered with food particles and bacteria. Many kinds of bacteria dwell in our mouth, making their living, for example, from the sugar we eat. We are not yet small enough to see these microbes. However, they make it painfully clear to nearly all of us that they really exist. These microbes produce acidic waste that can make holes in our teeth, causing us to suffer from a toothache.

and microbes

Some kinds of microbes peacefully coexist with the body cells. For example, some bacteria living in the intestine help us to digest food. Others, such as the bacteria causing cavities, can cause trouble. However, some microbes make their living by destroying body cells. They are the germs that cause infectious diseases. Outside the body, we can easily cope with microbes. Soap kills most viruses. When we boil water, the water becomes sterilized—the bacteria die.

Microbes intrude our body in different ways. Once they have entered, we cannot fight them with soap or hot water. We need other weapons—the cells of our immune system. Billions of freely floating defender cells keep our body free from intruders. They detect the enemy at any place in the body, and they have answers for every trick. These cells multiply as fast as they can when they encounter a microbe.

The immune system works constantly, each second of our lives. When things go fine, we do not notice its efforts. However, when too many microbes are intruding, we become sick.

A cold might be unpleasant, but it is harmless. After a few days, the immune cells have won their fight against the microbes. However, it hasn't happened yet!

What happens in
our *nose?*

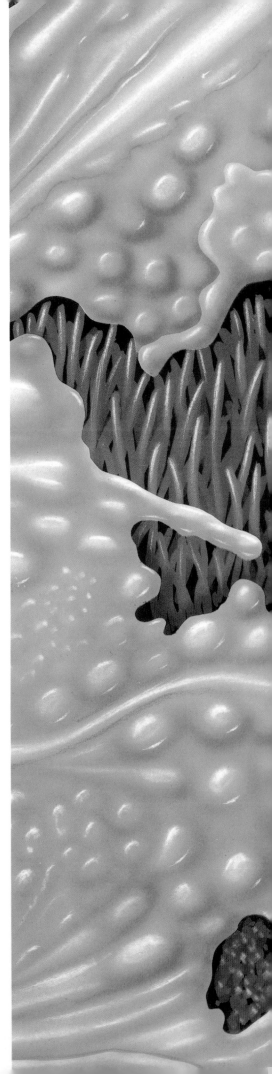

Here we are. See how infected nose cells look!

This carpet consists of mucous cells, which line the nose, mouth, and respiratory or breathing system. Their mucus is sticky enough to get hold of intruders. It also contains protective substances that destroy microbes. Fine hairs are in constant movement, and they push away microbes, dust, pollen, and dirt. When we blow our nose, we shove everything out. Coughing or clearing our throat loosens mucus and stuck microbes, and a sharp stream of air transports them up to the mouth. We swallow it all, and the stomach liquids kill and dissolve the intruders.

Whether you like it or not, mucus is one of the body's best weapons for getting rid of tiny germs. Mucus is produced by cells at risk of being attacked by microbes. When your nose is stuffed or runny, the cells in your nose are trying to protect themselves and our body.

However, this nose is stuffed because it is producing quite a lot of mucus. In spite of this, cold viruses have succeeded in sneaking through the mucous layer down to the cells. Viruses have infected many of them, as we can see from the area where the fine hairs are destroyed. Such inflammations occur at places where immune cells and microbes are fighting.

Why do we sneeze, Professor?
Intruding alien particles disturb the cells. Sneezing is a desperate attempt to expel the intruders and to clear the nose. It helps remove dust and pollen but doesn't work well against viruses. Most viruses remain. When we sneeze, we spray out germs into the air where they have a good chance of being inhaled by other people and infecting their nose cells.

How do cold viruses damage our cells?
Viruses transform healthy body cells into cells that do nothing else but produce fresh viruses. Sounds amazing, doesn't it? Well, once we have investigated the way cells—and viruses— function, we might be even more puzzled by their clever tricks. First, though, I suggest we take a closer look at the ways our skin tries to protect us against the minute invaders in the first place. The fewer microbes that come in, the better for us.

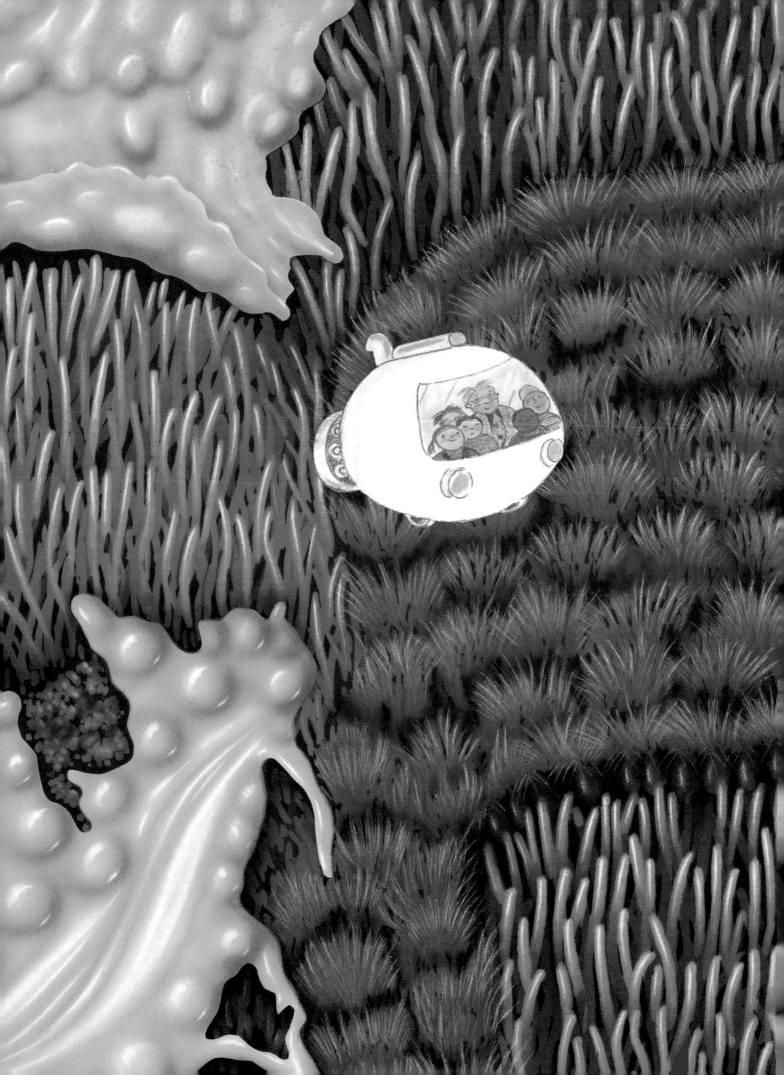

Body *fortress*

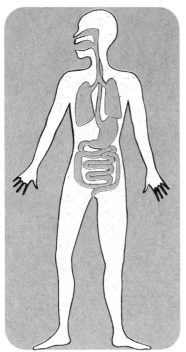

The skin that covers our body from outside is hard, dry, and has a top layer of dead cells.

The mucous skin that lines the inside of the body tube is soft and moist. It must, at the same time, be able to take in useful particles and fend off germs, dust, and other alien particles.

Just imagine your body as a kind of tube with very thick walls. Various substances and particles constantly wander into it, such as air, drinks, and food. Our cells take out the useful substances and use them to produce energy and as building materials. The waste matter and other useless things are removed.

Our skin has to protect us from the inside and from the outside. So, it has a quite simple task. On the inside of the tube, the skin is moist and soft, while the outside skin is rather dry and hard. The outer layer consists of hardened, dead skin cells acting as a protective armor. Viruses attack only living cells, so they are helpless in the face of these dead outer skin cells. Normally, bacteria cannot penetrate healthy skin.

Why is the inside skin moist, Gene?

It has a more complicated job than the outside skin. The mucous skin cells lining the body tube from the inside must fend off intruders. At the same time, they have to let in useful things. In the mouth, for example, cells sense flavor particles, which add taste to the food we eat. Within the nose, cells have to sense aroma particles. Within the lung, cells take up oxygen, and within the intestine, nutrients. So inside skin cells must know the difference between useful, useless, and dangerous things. They are much more sensitive than the outside skin cells. In spite of the antimicrobe mucus they produce, cells in the nose are quite at risk of being attacked by invaders.

The inside skin has to fend off all kinds of microbes and dust and dirt and tiny particles we constantly breathe in. Here we can see a small dust particle partly covered with slime. Soon it will be moved out of the body.

Now we have shrunk so small that we can even see these green bacteria—a few of many millions that live

on the top layer of our skin. Normally they won't do us any harm—as long as they do not penetrate the skin to get inside the body and reach the blood stream.

Can bacteria creep through our ears?

Well, they certainly would if they only could. However, our ears are protected by earwax, which contains substances that kill microbes. The same thing applies to tears, mucus, and sebum grease on the skin. Furthermore, all these liquids flow from the inside to the outside, constantly washing off would-be invaders.

How *cells* work

We've been talking about mucous cells, but, of course, hundreds of different kinds of cells build the various parts of our body. All of us—each and every living creature—consist of cells. They form the smallest units of life, minute pieces acting together to build an organism. Epithelial cells, for example, form a tight tissue lining the mouth, nose, lungs, throat, stomach, and intestine. Muscle cells form muscles that move our bones, which consist of bone cells. Nerve cells direct muscles' movement. There is a continuous interplay between cells and substances produced by cells, and that makes an organism function. Such an organism needs protection. Billions of immune cells drift in the blood and keep the body clean. We'll encounter them shortly. For now, this is how a cell looks from inside.

What kind of cell is this?

This one could be a mucous cell in the nose. Most cells work in similar ways. Their main task is to produce proteins. These are complex molecules used as building blocks of cells, as messengers organizing the cooperation of cells, and as triggers for chemical reactions. Nearly all cells consist of the same parts—organelles or cell organs. For example, cells have a skin called a membrane, an inner skeleton, protein factories called ribosomes, power stations called mitochondria, and a core called a nucleus. The nucleus is where the plans are stored that tell the cell exactly what to do. These plans are the genes—recipes for making proteins. There are 100,000 kinds of proteins at work in our body, doing 100,000 different jobs.

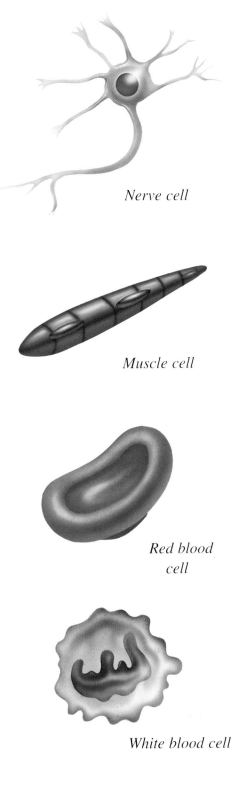

Nerve cell

Muscle cell

Red blood cell

White blood cell

Skin cell

Why can't we see genes here, Gene?

Genes are written on an extremely thin thread called DNA, which is stored in the nucleus. We'll have a closer look at genes later. Of course, the cell can read those recipes. What a cell actually does—and what kinds of proteins it makes—depends on which genes are switched on in that particular cell. For example, a mucous cell produces, among other things, proteins that form mucus.

It seems that the genes are a good starting point for someone or something that wishes to change the behavior of a cell.

Small, smaller, smallest

Ten micrometer

A white blood cell

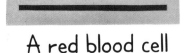

A red blood cell

Bacteria

Viruses

Viruses are really tiny. On the left we see how tiny they are compared with body cells and bacteria. By the way, those white lines represent a thousandth of a millimeter, or ten micrometers. Even when enlarged, a virus appears only as a tiny dot. The smallest virus ever detected is the polio virus. It would take 16 million of them to cover one millimeter.

Compared with a virus, a bacterium is quite huge. It takes more than one thousand bacteria laid next to each other to cover one millimeter. These are body cells—a red blood cell and, a bit larger, a white blood cell, one of our very brave immune cells.

How can such a tiny virus stay alive, Gene?

Actually, a virus doesn't live in the way bacteria, elephants, and humans live. A bacterium, for example, has all the bits and pieces a cell needs to make a living. Creatures like ourselves are alive thanks to the living cells we consist of. A virus, however, only consists of a cover of proteins with the viral genes inside. That's all. Viruses don't need food because they don't need energy. They do not eat, they do not move by themselves. They must be taken into a host cell in order to reproduce.

What is a host cell?

Cells infected by a virus are called host cells, although they are quite involuntary hosts. Some kinds of viruses infect host cells in animals (including humans). Other viruses concentrate on plant cells and even bacteria. Viruses are specialists. A cold virus, for example, wants to invade only the mucous cells in the respiratory systems of humans.

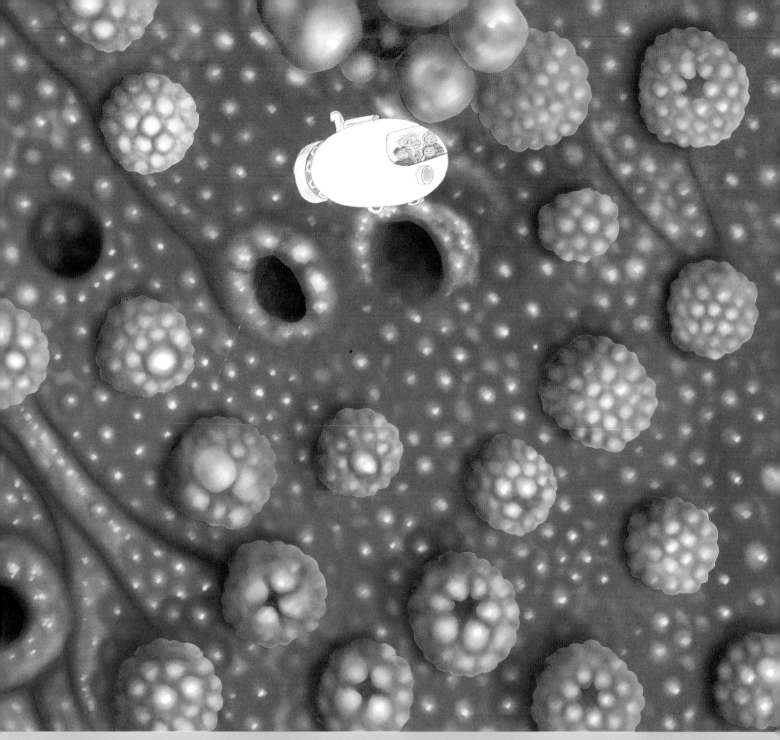

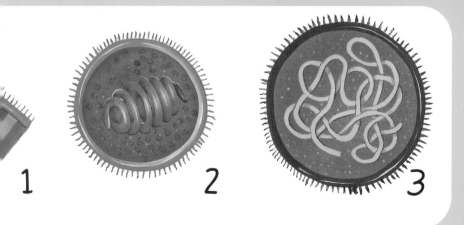

This is the membrane of one lip skin cell infected by herpes viruses. The viruses cause small bladders to form. Lots of fresh herpes viruses come out of these bladders.

There are many kinds of viruses, and all of them have a specific shape. Here we can see:

(1) A rabies virus,
(2) A flu virus,
(3) Polio viruses.

1

2

3

How a *virus* infects a cell

Normally, cells are protected by their membranes, and only useful things such as nutrients and messenger substances can penetrate them. This is because there are special kinds of proteins imbedded in the cell membrane. They act as doorkeepers and allow only those things in that have the right key—the right structures on their surfaces. The protein key and keyhold system allows cells to stick together in case neighboring cells of the same kind have the same keys and keyholes in their membranes. Because of that, mucous cells, or skin cells, or muscle cells, form a stable tissue.

So, how does a virus get into a cell?

It has the right key to it. So the virus can attach to proteins forming the keyhole on the membrane of the host cell. Then part of the cell membrane flips in, taking inside the attached virus. Of course, the doorkeeper doesn't mean to let the cold virus in. It only happens because the virus has the right matching protein structures on its covering surface.

Why are our colds only in the nose and throat?

Different kinds of cells have differently shaped keyhole proteins on their membrane. A cold virus, for example, only has the key to the cells that line the nose and throat, but not for other kinds of cells. Other viruses have different keys and therefore invade other cells. Different kinds of viruses therefore produce different diseases.

Once a virus has found an appropriate cell, it attaches to it and the host cell takes the invader inside. This is the first part of infection. Now the virus genes are inside the host cell. That is the beginning of the end of the cell. In the course of the attack, some kinds of viruses can

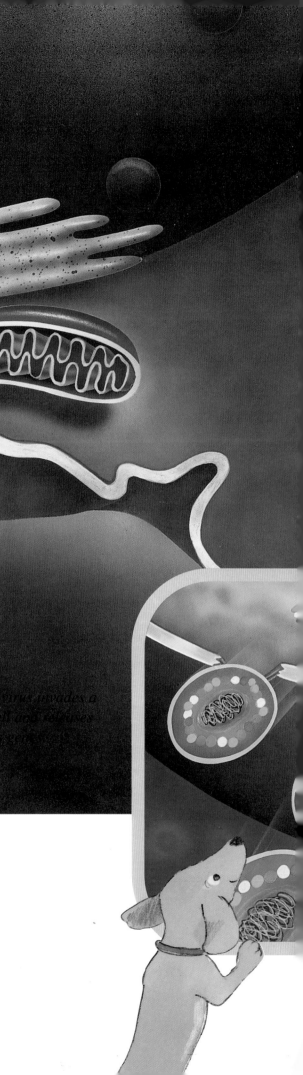

A virus invades a cell and releases...

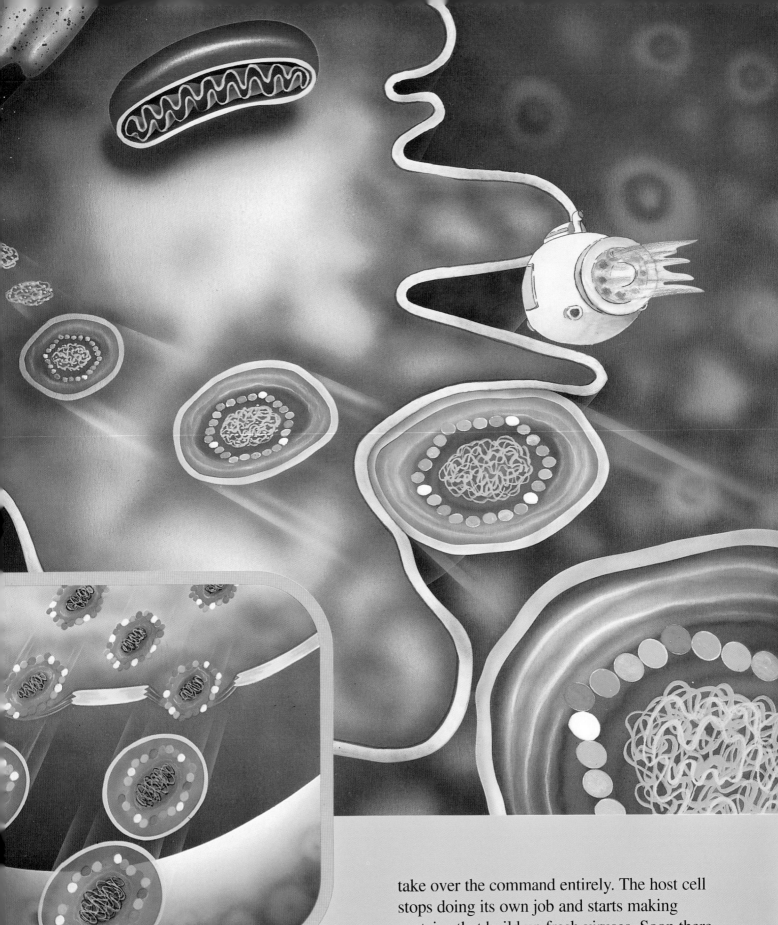

A few days later, lots of fresh viruses will swarm out.

take over the command entirely. The host cell stops doing its own job and starts making proteins that build up fresh viruses. Soon there are masses of fresh viruses bursting out of an infected, dying host cell. They swarm out, ready to infect more cells.

Why a *virus* can change a cell

Do viruses and host cells have the same genes?

Of course not. The genes are different, but their building blocks are the same. Let's have a look at how genes work.

Within body cells, genes are packed in parcels called chromosomes. In many cells, including all human cells, the chromosomes are within the cell nucleus. Chromosomes are made of an incredibly thin threadlike substance called DNA (deoxyribonucleic acid). DNA consists of two strands of building blocks wound around each other. When unraveled, DNA looks like a ladder. The building blocks— the rungs of the ladder—are called nucleotides. They come in four *flavors* and are lined up in a specific sequence. All the information about how to assemble proteins is stored on DNA in the nucleotide sequence. Our genes are the parts of DNA that contain the recipes for making proteins. A mucous cell, for example, learns from its genes how to make mucus.

In a virus, which has no nucleus, DNA is just curled up like noodles on a plate. However, if we were to unravel the viral DNA, it would look very much like ours—like a twisted ladder. Both human and viral DNA are made from the same kinds of nucleotide building blocks.

In viral DNA, however, those same nucleotide building blocks are lined up in completely different sequences than in the host DNA. Therefore, a virus has different genes, different recipes for making proteins. Once these alien virus genes get into the host cell, the host is infected. It starts doing what the virus genes tell it to do—to make fresh viruses.

Which diseases do viruses cause?

They can cause colds, influenza, and measles but also very dangerous diseases like polio, smallpox, and rabies. However, not all viruses make us ill immediately. Some viruses take their time. They invade their host cell and then wait for their chance to begin working when the immune system is busy fighting at other places. Herpes viruses, for example, which cause the herpes blisters on lips once in a while, are sleeping viruses that some people carry around their whole lifetime. Some

A body cell and a virus both contain genes stored in DNA.

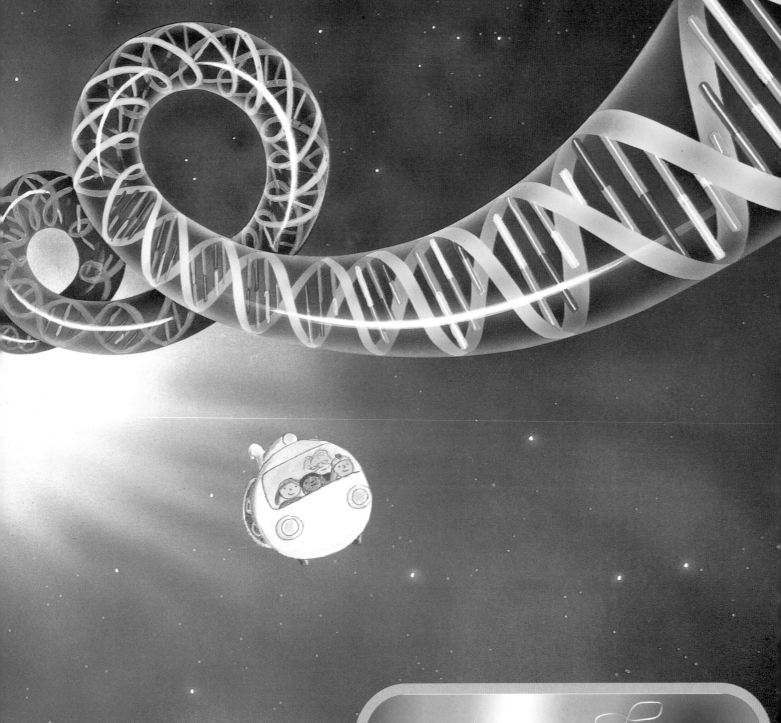

...iruses cause warts and let their host cell stay
...live—making the host cell divide more and more.

...By the way, many kinds of viruses don't do any
...arm to humans. For example, there is a virus
...hat infects only the mucous cells of a cat. Even
...when an infected cat sneezes at us, spraying
...millions of viruses, we humans won't be infected.
...imilarly, viruses infecting the roots of carrots
...re harmless for us and for cats. However, cold
...iruses not only infect people but also ducks and
...igs. The viruses' key matches specific cells in
...igs, ducks, and humans.

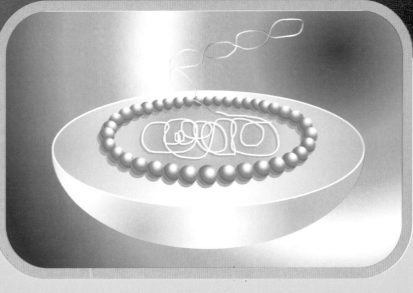

A virus and its DNA.

Bacteria, fungi, and the *nasty* others

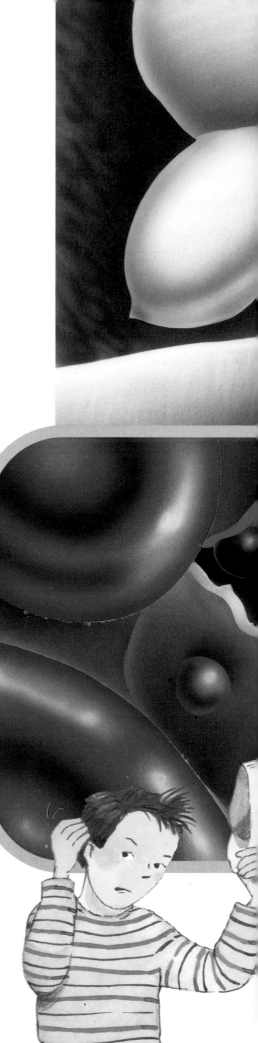

Bacteria work differently. Bacteria are alive like animals and plants, although they consist of a single cell. Bacteria form an important part of nature. For example, they break down dead animals and plants and transform them into fertile soil. Most kinds of bacteria living in our mouth, nose, and intestines are harmless or even useful. Intestinal bacteria, for example, help us to digest food. Can you imagine that you have nearly one pound of bacteria in your gut? Other kinds, however, are quite dangerous and can make us ill.

How can bacteria cause diseases, Gene?

Since they are living creatures, bacteria take substances from their environment and use them to produce energy and to make the things they consist of. In doing so, they produce lots of waste, which can poison us or destroy, for example, our teeth. Bacteria that feed from sugar cause dental cavities. Their acidic waste products dissolve the hard enamel covering the teeth and make holes in it.

Bacteria make us sick in many ways. Some overpower our cells and eat them. Others keep our cells from doing their job properly. The cholera bacterium, for example, makes the cells lining the intestines lose water, causing a terrible and deadly diarrhea. More common and less dangerous problems caused by bacteria are ear infections, stomach pains, and even pimples.

How dangerous is one single bacterium?

The real problem with one bacterium is that it multiplies at an incredible speed, provided there is enough for it to eat. Unfortunately, our body provides sheer endless quantities of food. Some kinds of bacteria can divide every 20 minutes. A couple of hours later there are millions and billions of them. This is why even a single bacterium could be hazardous—it can multiply and multiply.

In addition to bacteria, other kinds of germs also exist. Fungi is an example, but not those that grow in the woods. We are talking about things like athlete's foot and other nasty fungi that can thrive in the lungs or in the intestines. Here we are about to cross through a colony of fungi.

Parasites also try to make their living within us by feeding from the food our body supplies. Parasites are living organisms that come in many sizes and kinds, for example as worms or as really dangerous creatures such as malaria germs. See this red blood cell? It is about to be destroyed by malaria germs.

Attention, *invaders!*

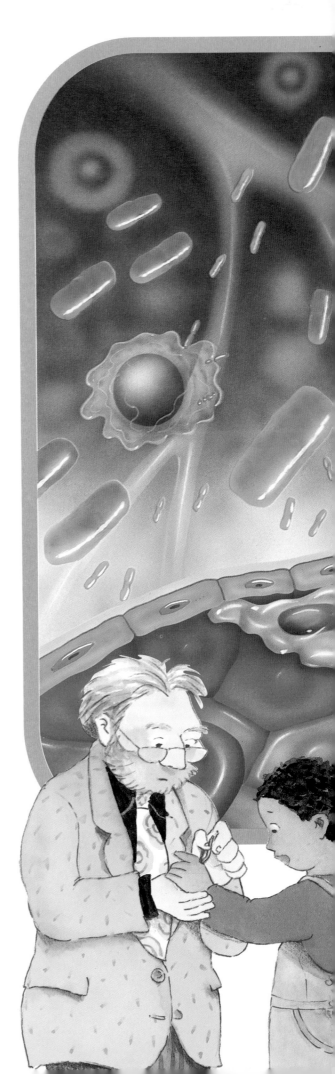

Outside the body, most microbes are harmless. Cold viruses stuck on the outer skin are helpless and so are bacteria such as *Escherichia coli*. They become dangerous only when they get inside the body. Their favorite way to get there is hitchhiking on food, especially on meat. For a cold virus, things are easier. We breathe it in or transport it by putting our fingers inside our nose or mouth.

Other microbes, however, may gain direct access to the body's bloodstream—a wound, a small cut, or a splinter in the finger might be enough. All of a sudden, the door into the body is wide open. Invaders can take the chance to swim the bloodstream and find their favorite places.

See that splinter? It has broken the body's first defense line, the skin. The body is now in a state of alarm. Green bacteria are streaming in. See what happens now.

Hurt cells sound the alarm. They send out messenger substances to the cells of the blood vessels. The blood vessels widen and more blood streams into the hurt cell tissue. The skin turns red and warms, because warmer blood is coming from inside the body. Also, the connection between the blood vessel cells becomes looser. More liquid can flow from the vessels, and the area becomes swollen.

Messenger substances from the hurt cells also lure macrophages and other defender cells swimming in the bloodstream to the wound. These messengers make the blood vessel cells sticky in order to catch the defender cells and make them quickly reach the inflamed cell tissue. The defender cells squeeze their way out of the blood and through the tiny gaps between the blood vessel cells. Then they start doing their job—eating microbes and damaged cells. Now we know why macrophage means *big eater*.

Defenders
at work!

We each have more than a trillion defender cells in us—hundreds of times more than the population of our planet. They are white blood cells, more clear than white, and are made in the bone marrow and thymus. They usually drift with the blood and with the lymph. The lymph is a clear liquid that flows in its own system of vessels throughout the body.

The immune system consists of many kinds of white blood cells. Some are designed to fight and eat any invaders.

One kind is called granulocytes. The yellow milky liquid seen in a wound is called pus. It comes from the remains of microbes killed by granulocytes and from our own cells that lost their lives during the battle.

Macrophages, those *big eaters,* are larger white blood cells than the granulocytes. They roam through the body and assemble at infected areas where they find edible things such as microbes. They also eat damaged body cells, cells at the edge of a wound or worn-out red blood cells, for example.

How can cells eat bacteria, Gene?

Well, here we see it. A macrophage, a defender cell, is about to finish quite a few bacteria. First, the big eater brings them closer with its pseudopods. These are *false feet.* The defender cell now embraces one invader after another in a deadly hug, brings them inside, and breaks them down into eatable particles.

Antibodies appear on the battlefield, along with blood and lymph. Antibodies are tiny, tiny cell-made protein particles that stick to the invading microbes and form lumps on their surfaces. They increase the appetite of the macrophages, which then use all their efforts to eat the invaders marked by antibodies.

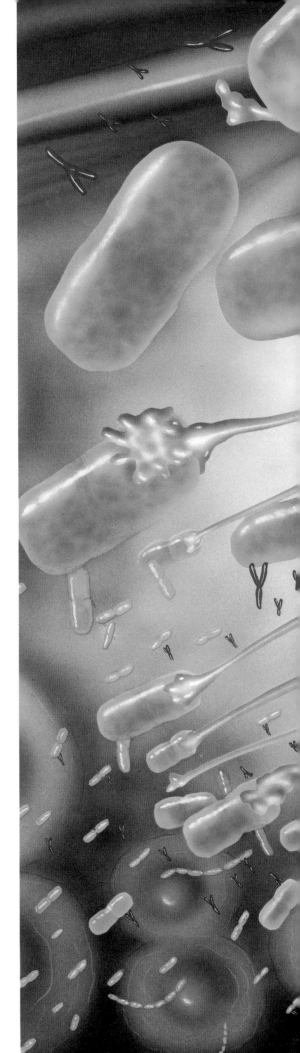

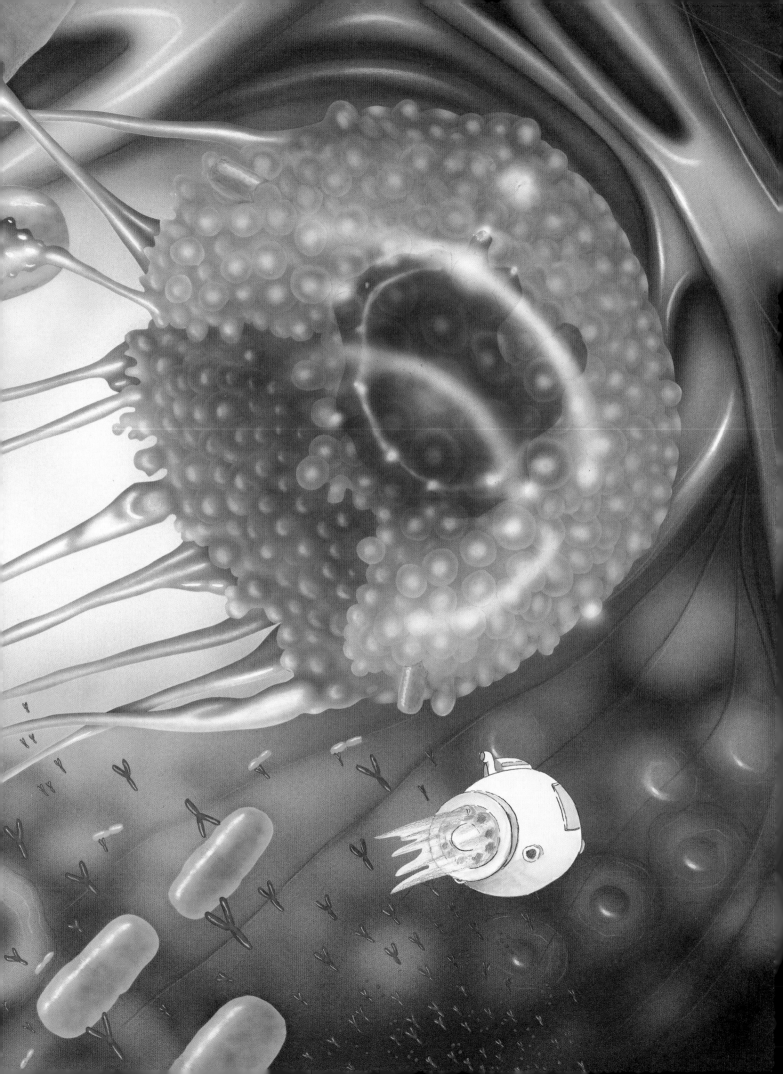

Friend or *foe* ?

Macrophages and granulocytes may fight as bravely as possible, but they alone are not capable of finishing off all microbes and stopping the invasion. Even if more and more macrophages were sent in, too many microbes might still sneak away and multiply at other parts of the body. They might even hide in other cells of the body. So the body badly needs other kinds of fighters and weapons, those specially built and trained for fighting only the very kinds of invaders infecting the body. This is the task of T cells (some of which are called killer T cells), of B cells, of antibodies, and of chemical bombs called complement. B and T cells are part of the so-called specific immune system—they recognize, hunt, and destroy specific germs.

Before it can take any action, the immune cells need to know exactly which kinds of microbes have invaded the body. Only then can the specialized defender forces be built and multiplied. This may take quite a while.

How do defender cells find out what is a friend and what is a foe?

Our body cells have identity marks floating on their surface by which they can be identified. In a way, this works like presenting a passport containing a photograph of the owner. Of course, cells do not show photos. Instead, they present to defender cells pieces of their proteins attached to special proteins in their cell membranes. Defender cells use these proteins to recognize body cells as friends and leave them in peace.

Bacteria, viruses, and other alien particles, however, do not have such *friendly* marks. Their surface proteins are alien. So they are dealt with as enemies.

This is how different kinds of defender cells and weapons work together to fight bacteria.

1

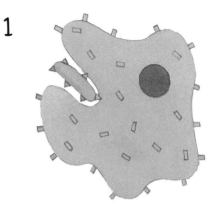

(1) A defender cell, a macrophage, is about to embrace and eat a green bacterium. The rectangles inside the defender cell are tiny identity proteins, also shown on its membrane. The bacterium has different identity marks, here shown as triangles on the surface. These proteins are alien to our body. We call them antigens.

2

(2) Now the macrophage has eaten the bacterium. Some bacterial antigens— the triangles—are presented at the surface of the macrophage, together with the macrophage's own identity proteins—the rectangles.

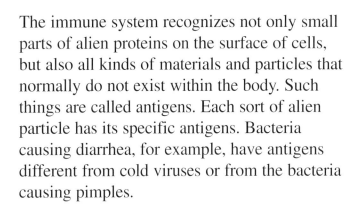

The immune system recognizes not only small parts of alien proteins on the surface of cells, but also all kinds of materials and particles that normally do not exist within the body. Such things are called antigens. Each sort of alien particle has its specific antigens. Bacteria causing diarrhea, for example, have antigens different from cold viruses or from the bacteria causing pimples.

What is an antigen, Professor?

An antigen is something the body considers foreign. It triggers the production of antibodies, which are essential weapons against germs. So let's have a closer look at them.

3

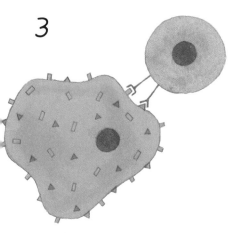

4

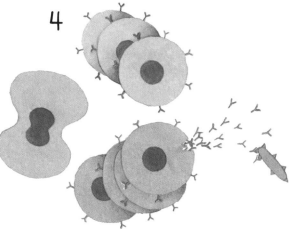

(3) Now a very special defender cell, a white blood cell called B cell, comes along. It finds out that the macrophage carries not only its own friendly identity marks but also alien antigens—the remains of the bacterium. The B cell becomes alarmed. It knows now that bacteria of a specific kind have invaded the body.

(4) Immediately, the B cell starts dividing to form different kinds of defender cells. One kind, called plasma cells, start producing antibodies. The other kind, called memory cells, will remember what kind of germ has invaded the body. Lots and lots of antibodies produced by lots and lots of plasma cells swarm into the bloodstream and lymph.

They cling to the antigens of more bacteria in order to mark them as aliens. Explosive complement proteins will attach to the bacteria and destroy them. Macrophages come along to deal with the rest of the nasty invaders.

Antibodies!

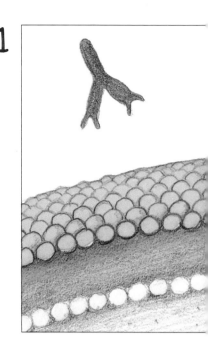

Antibodies are proteins that come in the shape of a Y. There are lots and lots of different models drifting in the blood and lymph. Each antibody model matches a specific antigen—just like a key fits only one specific keyhole. Once an antibody encounters a bacterium or a virus carrying its precise antigen, both of the antibody's arms stick tightly to the antigen.

Why don't we make enough antibodies in advance?

The immune system cannot work ahead. It is impossible to know which kinds of disease may threaten us tomorrow or next year. It is also impossible to have a reserve of many billions of antibodies for each possible kind of enemy. The immune system works on the basis of demand. Only after microbes have invaded us can our body find out which kinds of microbes it has to deal with. After that, it can produce billions of tailor-made antibodies.

How does an antibody find its antigen?

Antibodies cannot move by themselves. They float with the blood and lymph. Sooner or later they encounter a microbe with the matching antigen and stick to it.

What exactly do antibodies do?

They help the defender cells in many ways. For example, antibodies attached to bacteria make them more tasty to macrophages. They also keep microbes from functioning properly. Some kinds of antibodies mark bacteria for the attachment of a sort of biological dynamite called complement.

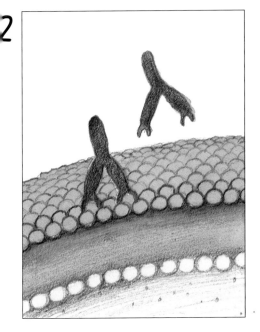

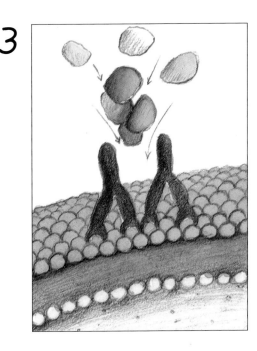

This is how antibodies and complement proteins work together:

(1) An antibody has found a bacterium.

(2) It clings to the bacterium's membrane.

(3) Now complement proteins start assembling on the bound antibody.

(4) The proteins, one after another, settle at the very spot that has been marked by the antibody.

(5) The complement proteins start forming a hole in the membrane of the bacterium.

(6) Finally, a pore is formed by the complement proteins. The interior of the bacterium leaks out, leaving behind a dead invader.

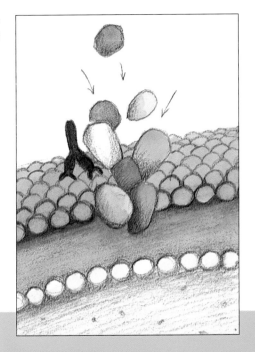

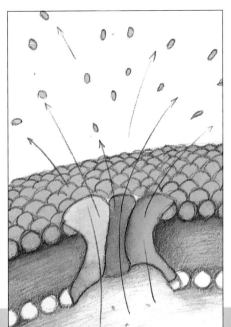

Explosives in our blood, Gene?

In a way, yes, explosives are in our blood. The complement is a system of nine different proteins with remarkable powers. They drift, quite harmlessly at first, in the bloodstream. However, once they encounter a bacterium coated with antibodies, they cling to the surface of the invader one after the other. Only when all nine complement proteins sit on the cell membrane does the explosive detonate. The complement tears a hole in the membrane, and the bacterium bursts and dies. You see, complement is a powerful weapon system. Therefore, it must detonate only on cells marked by antibodies—on alien cells, such as bacteria.

More
antibodies!

All the various kinds of immune cells descend from stem cells that have their home in the bone marrow. Their offspring become either macrophages (those greedy eaters) and granulocytes or defender cells, specialized in fighting specific germs. B lymphocytes, or B cells, are those defender cells; they make antibodies.

How do B cells know which antibody to make?

Well, first of all, the number of different models of B cells is huge. Each model can make only one kind of antibody. Samples of all these different B cell models constantly drift through the blood and the lymph. They just wait for the right antigen, for their antigen, to come along and to order a few billion antibodies.

Now if a macrophage has swallowed a bacterium, it brings up to its surface the antigens of the gobbled invader and starts roaming around. The macrophage presents the antigen to all the B cells it meets. Most B cells are not interested in these particular antigens. At last, the macrophage encounters the right kind of B cell. And this B cell, recognizing the antigen, is alarmed. It starts to divide, and its offspring go on dividing again and again. They become so-called plasma cells that do nothing but produce and send out incredible amounts of antibodies. Each plasma cell makes 30 antibodies per second. Soon billions of fresh, tailored-to-fit antibodies swarm out. They detect and attach themselves to their corresponding microbes.

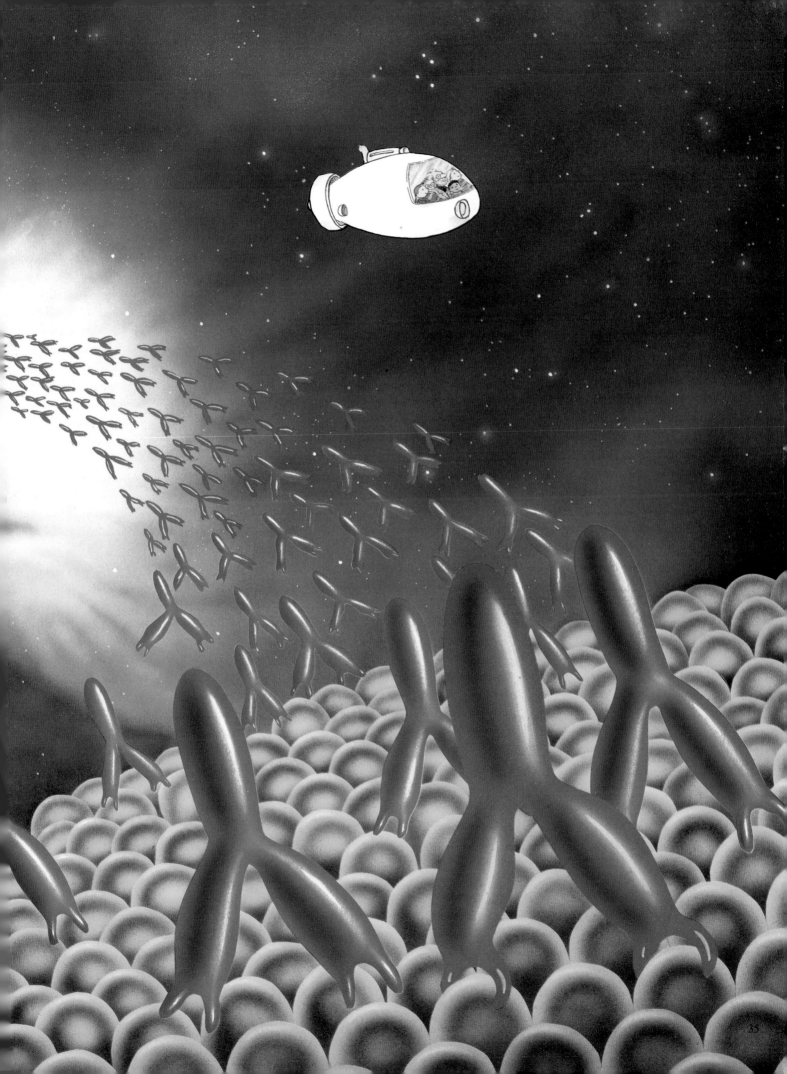

The attack of the
killer T cells

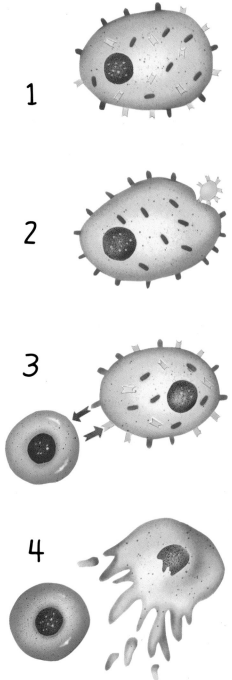

1

2

3

4

Viruses are not as easily detected as bacteria. They know a simple trick to protect themselves from antibodies— they hide inside body cells. All the immune cells can do is prevent the viruses from multiplying inside the host. This is the job of the killer T cells. They detect and poison infected body cells. Just don't forget that they are the heroes, not the villains.

Where do killer T cells come from?

They are another kind of white blood cell. T stands for the thymus gland under the breast bone. Samples of many thousands of different killer T cell models are made and tested all the time in the thymus. Some killer T cell models would turn out to be mad killers keen on poisoning healthy body cells. They must be destroyed immediately. The other models that react only to alien antigens are sent out into the bloodstream and lymph stream. Each model will watch out for cells infected by a specific virus. Once a killer T cell detects such an infected cell, the killer T cell starts multiplying to make more of itself. Some days later, zillions of killer T cells stream out into the blood and lymph. Throughout the body, they destroy all cells infected by this specific virus.

This is how killer T cells clear our body from cells infected by a virus.

(1) The big cell is a skin cell in the nose. The cell contains specific proteins that also show on the surface. They are identity marks that make clear to the defender cells that this cell belongs to the body. The tiny yellow cold virus that is about to come in also has identity marks—the sharp yellow things—but alien ones.

(2) The body cell is infected and ruined beyond repair. However, before dying, it has enough time to bring pieces of the cold virus, the viral antigens, up to its membrane and display them there. This means: Look here, killer T cells, I am infected! Destroy me quickly before I make fresh viruses!

(3) And here the killer comes. It is trained to detect body cells that show both their own identity marks (the pink round ones) and the viral antigens (the yellow, sharp ones). The killer T cell attaches using its unique receptor model that works like a keyhole. The T cell receptors match with a bipartite key on the infected cell—one part is the viral antigen, the other part is the body cell's own identity marks. Now the killer T cell understands that the body cell is infected.

(4) The killer T cell will inject its poison and destroy the infected body cell along with its deadly load of viruses. Now the T cell drifts off to look for other cells infected by this specific virus.

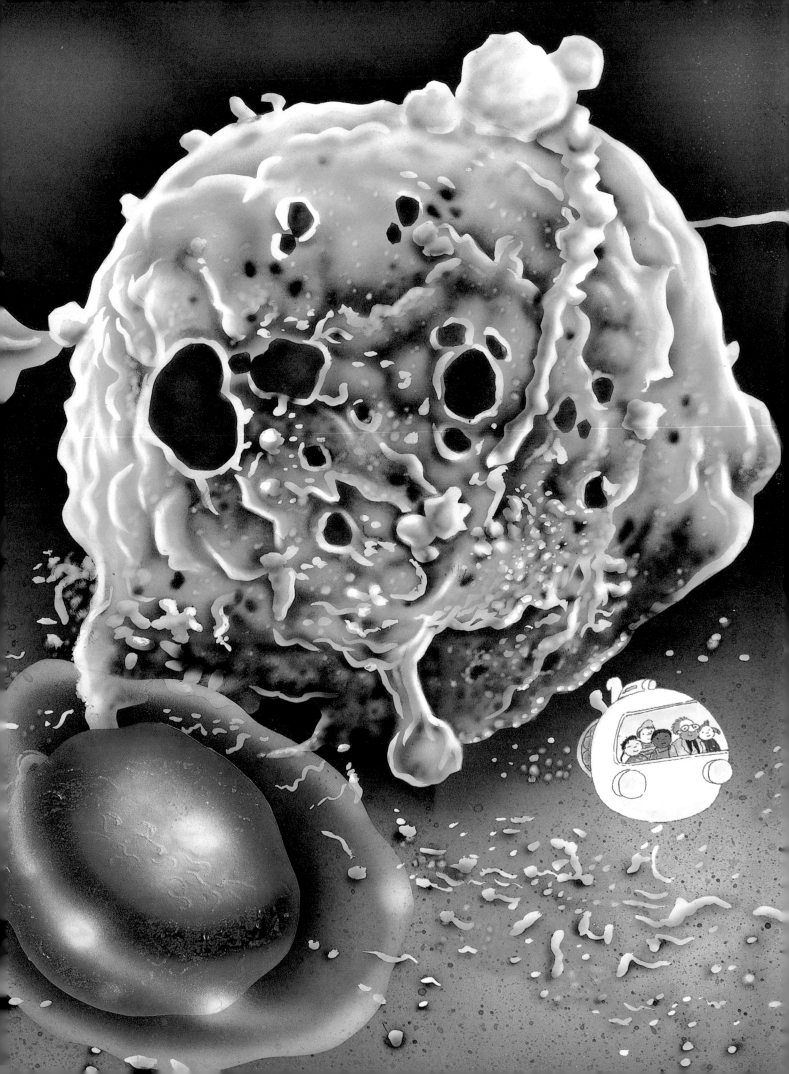

Feeling *ill*

Microbes that have invaded the body multiply as fast as they can. At the same time, the immune cells start producing really large numbers of antibodies and killer T cells tailored for this very kind of invader. It takes a while until enough of them appear to finish off the microbes. Meanwhile, we feel ill.

However, it is not the germs that cause fever and sickness. It is the immune cells that send out messenger substances telling the body to feel weak, sick, and tired: Go to bed and rest! Don't run around! Don't waste the energy that the defenders need now!

Why do we have fever?

Higher body temperature, or fever, is something that the body does to become healthy as soon as possible. Higher body temperatures speed up production of antibodies and killer T cells. After a few days, they have done their job and destroyed all microbes. We feel fine again.

As we've seen, it is a difficult task first to identify the invader and then to build from scratch the weapon systems against this particular germ. However, there is a short cut.

Why do we get measles only once, Gene?

Once the immune cells have fought off a specific microbe, such as the measles virus, some defender cells called memory cells remain on alert in the blood—just in case these germs come back. Now, if the measles virus tries it again, the defenders know exactly what to do. Without delay, masses of the right antibodies are produced. They overthrow the invaders before the microbes even begin multiplying.

If you ever had measles, it is unlikely you would get it a second time—even when you visit a friend suffering from measles. Your defender cells know all too well how to fight the measles virus: you are immune.

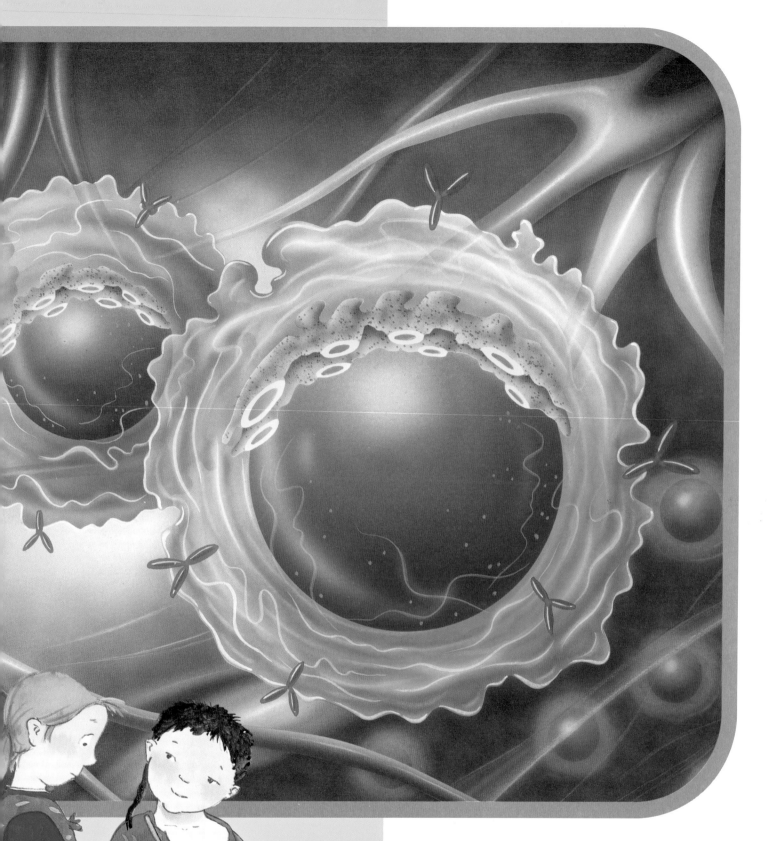

Take a good look at these memory cells. They are white blood cells that remember germs that have infected us before.

Why do we catch colds over again?

The germ changes its outfit. For example, this year's cold virus looks different from last year's cold virus. The memory cells remember last year's model but have never met this year's.

39

Supporting *our*

When it comes to fighting bacteria, antibiotics are helpful. These medicines kill bacteria. Penicillin, for example, which is produced by a fungus, prevents bacteria from building a fresh cell wall after dividing. See here a colony of bacteria suffering from a penicillin disease. However, since antibiotics also kill useful bacteria, we should take such medicine only when needed.

What about fighting viruses with antibiotics?

Viruses don't live in the way bacteria live, so antibiotics do them no harm. Fighting viruses depends completely on the immune cells. We can only support immune cells in their struggle—by receiving a vaccination.

What happens at a vaccination?

This is a clever trick to prepare immune cells for a virus attack in advance. A vaccination introduces weakened or killed bacteria or viruses into the body. To the immune cells, vaccines look like the real and dangerous germs because they have the same antigens. The immune system starts producing the proper antibodies. The memory cells remember the patterns of the invaders. If the microbes attack, the defender cells can deal with them immediately. We stay immune as long as the memory cells stay alive, maybe for many years. However, antibodies can also be taken from the blood of sheep or horses and given to a person. These antibodies instantly protect an infected person who, for example, was bitten by a dog suspected of carrying the dangerous rabies virus. Since these antibodies are not made by the person, the immune system does not create memory cells of them. These antibodies soon wear off.

Gene, can I have a vaccination against—atchooo!

Bless you. Unfortunately, you cannot be vaccinated for a cold. As you know, cold viruses constantly change their outfit. Don't worry, though. Your defender cells have already identified these nasty invaders. They will soon destroy the virus. I think the immune system deserves our appreciation.

So don't be too impatient. Your body's heroes really fight bravely!

Now let's quickly get out of the nose and return to our normal size before . . . atchoo!!

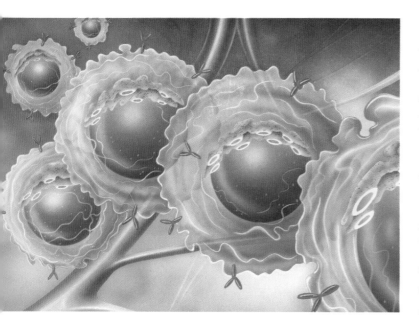

b) is correct. When a virus enters a healthy body cell it brings along its own genes. Instead of doing its regular job, the cell starts to produce fresh viruses.

Bacteria can cause

a) influenza
b) measles
c) rabies
d) cavities
e) cholera

d) and e) are correct. Dental cavities are caused by the waste produced by bacteria that eat sugar. This is a very common bacterial disease and most people suffer from it. Compared to cholera, dental cavities are, of course, quite harmless. Influenza, measles, and rabies are caused by viruses, not bacteria.

Fever is caused by

a) bacteria
b) viruses
c) messenger substances produced by the body

c) is correct. When the body is seriously infected by viruses or bacteria, it turns up its inner heating. An elevated body temperature speeds up the healing processes.

What do you remember?

Substances that help the body fight against an infection caused by bacteria are called

a) antibodies
b) antigens
c) antibiotics

a) and c) are correct.

Compared to bacteria, viruses are

a) smaller
b) not alive
c) resistant to antibiotics

a), b) and c) are correct.

Viruses can make us ill by

a) eating up cells
b) infecting healthy cells and causing them to make more viruses
c) poisoning infected cells

Defender cells tell friends from foes by

a) their taste
b) their antigens
c) their behavior

b) is correct. Antigens are substances that normally are not found in the body. One the immune cells encounter antigens, plasma cells produce large amounts of antibodies and killer T cells attack infected cells.

The cells and weapons of the immune system deal with microbes by

a) eating them
b) poisoning them
c) making holes in them
d) expelling them

All answers are correct. Macrophages, for example, eat microbes. T cells poison cells infected by a virus. The chemical explosives of the complement system perforate bacteria marked by antibodies. And the tiny hairs of mucous cells wave and expel microbes that they capture in the slimy mucus.

A vaccination

a) fights infections with animal antibodies
b) makes us produce antibodies without feeling ill
c) uses dummy viruses

a), b) and c) are correct. During a passive vaccination, antibodies produced within animals are injected into the bloodstream to give quick protection against a virus. Active vaccinations bring in harmless dummy viruses that make the body produce its own antibodies against a particular disease. And all this without making us feel ill!

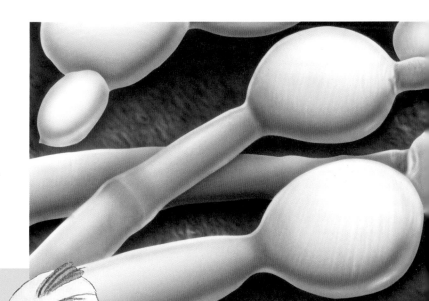

Glossary

Antibiotics chemicals that kill foreign cells such as bacteria.

Antibodies proteins that help kill bacteria.

Antigen any particle (or part of a particle) the body considers foreign. The word antigen is short for antibody-generator. Soon after an antigen appears in the blood, the right antibodies are made.

B cells (or B lymphocytes) white blood cells specialized in making antibodies.

Bacteria tiny organisms that consist of only one cell. They are smaller than body cells. They come in thousands of kinds. Some are helpful to our body; some are dangerous and can cause diseases.

Bacterial diseases caused by bacteria that can poison our cells with bacterial waste, or keep them from functioning properly, or even kill body cells. Cholera, ear infections, and even pimples are caused by different kinds of bacteria.

Blood carries nutrients, oxygen, and many other substances to all parts of the body and removes the poisons and waste products. It also contains many different kinds of cells.

Bone marrow the innermost part of bones where fresh red and white blood cells are made.

Cells tiny living beings. About 200 different kinds of cells form the human body and help to keep it alive. Most of them are so small that you could lay 100 of them side by side in less than one-tenth of an inch.

Complement a system of different proteins designed to tear holes in a bacterium.

DNA short for deoxyribonucleic acid. Genes are lined up on these incredibly thin strings, shaped like twisted ladders, that are found in each cell.

Fungi germs that live in and on our body. They can, for example, cause athlete's foot and some dangerous diseases.

Genes sections on the DNA. Stored in the cell nucleus, they are the recipes for making proteins. Each gene consists of several thousand code words.

Herpes viruses infect skin cells and cause the skin to form small blisters.

Host cells body cells that have been infected by viruses.

Immune system fends off germs. It consists of billions of white blood cells and of the weapons they produce. They come in many different kinds and work together to keep our body free from intruders.

Inflammations occur at places where the cells of the immune system fight germs.

Killer T cells destroy body cells that are infected by viruses.

Lymph a clear liquid that flows in its own system of vessels throughout the body.

Macrophages ("big eaters") a kind of white blood cell. They patrol with the blood and eat microbes as well as damaged cells.

Membranes moist films with many different proteins stuck inside. The membrane keeps the cell shielded from the outside, makes cells stick together, and also receives and sends out signals.

Memory cells white blood cells that remember germs that have infected us before so the specific immune system can be built up without delay.

Microbes tiny germs such as bacteria, viruses, fungi, and parasites that live in and on other beings.

Mucus a liquid produced by specialized cells that contain substances that kill germs.

Parasites creatures that make their living by feeding from the food supplied by the host they live in or on.

Plasma cells white blood cells that produce antibodies.

Proteins the building blocks of cells and the tools that cells use for their many kinds of activities. They are made according to the recipes provided by the genes.

Skin fends off microbes and keeps them from invading the bloodstream.

Specific immune system consists of specialized defender cells such as B cells and T cells that only answer to the attack of a specific germ.

Thymus a gland under the breastbone where killer T cells are made and trained.

Vaccination prepares our immune cells for a possible attack of a specific kind of germ.

Viral diseases caused by a virus such as various flu viruses or the dangerous rabies virus.

Viruses incredibly tiny and sometimes really dangerous creatures that have no life on their own. They inject their own genes into body cells and turn healthy cells into sick cells that produce fresh viruses.

White blood cells come in many different kinds. They are constantly on the outlook for harmful germs that they destroy.

Index